YOUR KNOWLEDGE HAS VALUE

Bibliographic information published by the German National Library:

The German National Library lists this publication in the National Bibliography; detailed bibliographic data are available on the Internet at http://dnb.dnb.de .

Imprint:

Copyright © 2011 GRIN Verlag, Open Publishing GmbH
Print and binding: Books on Demand GmbH, Norderstedt Germany
ISBN: 9783668293069

This book at GRIN:

http://www.grin.com/en/e-book/339110/on-zeros-of-laguerre-polynomials

Ayo Odeniran

On Zeros Of Laguerre Polynomials

GRIN Publishing

GRIN - Your knowledge has value

Since its foundation in 1998, GRIN has specialized in publishing academic texts by students, college teachers and other academics as e-book and printed book. The website www.grin.com is an ideal platform for presenting term papers, final papers, scientific essays, dissertations and specialist books.

Visit us on the internet:

http://www.grin.com/

http://www.facebook.com/grincom

http://www.twitter.com/grin_com

ON ZEROS OF LAGUERRE POLYNOMIALS

BY

ODENIRAN Sunday Ayodeji

Chapter 1

Orthogonal polynomials

1.1 Introduction

Markov and Stieltjes in the late 19th century established the first results on the behaviour of the zeros of orthogonal polynomials, and this field of research has held the attention of theoreticians, functional and numerical analysts ever since then. Some of the tools and techniques developed to analyse different properties of the zeros are Markov's theorem on the monotonicity of zeros ([26]); Sturm's comparison theorem for the zeros of solutions of second order differential equations ([?]); Obrechkoff's theorem on Descartes' rule of sign ([30]); and the Wall-Wetzel theorem on eigenvalues of Jacobi matrices ([36]).

There is an extensive literature in the field of orthogonal polynomials which is particularly concerned with detailed investigation of zeros and related questions (e.g., [29], [34]). Most results on the zeros of the classical orthogonal polynomials are local in nature, such as inequalities, asymptotic expansions, and monotonicity properties. For instance, it is well known that if $\{p_n\}_{n=0}^1$ is any orthogonal sequence, then the zeros of p_n are real and simple and each open interval with endpoints at successive zeros of p_n contains exactly one zero of p_{n-1}; a property called the interlacing of zeros. Stieltjes (cf. [?], Theorem 3.3.3) extended this interlacing property by proving that if $m < n - 1$, provided p_m and p_n have no common zeros, there exist m open intervals, with endpoints at successive zeros of p_n, each of which contains exactly one zero of p_m. Beardon (cf. [5], Theorem 5) proved that one can say more, namely, for each $m < n - 1$, if p_m and p_n are co-prime, there exists a real polynomial S_{n-m-1} of degree $n - m - 1$ whose real simple zeros together with those of p_m, interlace with the zeros of p_n. The polynomials S_{n-m-1} are the dual polynomials introduced by de Boor and Saff in [6] or equivalently, the associated polynomials analysed by Vinet and Zhedanov in [24].

In recent years, authors including Ismail and Muldoon (cf. [14]), Krasikov (cf. [16]), Gupta and Muldoon (cf.), Segura , Ismail and Li (cf.), and Dimitrov and Nikolov (cf.) have developed interesting methods including the use of chain sequences and the derivation of inequalities for real-root polynomials to refine and improve upper and lower bounds for extreme zeros of classical orthogonal polynomials.

In this work, we make attempt to gather different results by many mathematicians who have made landmark achievements on studies on zeros of Laguerre polynomials. We discuss bounds for extreme zeros of some classical orthogonal polynomials, Bounds for zeros of the Laguerre polynomials where we quote the works of Ilia Krasikov [25], Limit relation for the complex zeros of Laguerre polynomials, On the Asymptotic Distribution of the Zeros of Laguerre Polynomials, Monotonicity of zeros of Laguerre polynomials, Zeros of linear combinations of

Laguerre polynomials from different sequences, Linear combinations of Laguerre polynomials of the same degree, Linear combinations of Laguerre polynomials of different degree and Convexity of the extreme zeros of Laguerre polynomials

1.2 Preliminaries

Polynomials A polynomial in a single indeterminate can be written in the form

$$a_n x^n + a_{n-1} x^{n-1} + ... + a_2 x^2 + a_1 x + a_0$$

where $a_0, ... a_n$ are numbers, or more generally elements of a ring, and x is a symbol which is called an indeterminate or, for historical reasons, a variable. The symbol x does not represent any value, although the usual (commutative, distributive) laws valid for arithmetic operations also apply to it. This can be expressed more concisely by using summation notation:

$$\sum_{i=0}^{n} a_i x^i$$

That is, a polynomial can either be zero or can be written as the sum of a finite number of non-zero terms. Each term consists of the product of a number - called the coefficient of the term - and a finite number of indeterminates, raised to integer powers. The exponent on an indeterminate in a term is called the degree of that indeterminate in that term; the degree of the term is the sum of the degrees of the indeterminates in that term, and the degree of a polynomial is the largest degree of any one term with nonzero coefficient. Since $x = x^1$, the degree of an indeterminate without a written exponent is one. A term and a polynomial with no indeterminates are called respectively a constant term and a constant polynomial; the degree of a constant term and of a nonzero constant polynomial is 0. The degree of the zero polynomial (which has no term) is not defined.

A polynomial function is a function that can be defined by evaluating a polynomial. A function f of one argument is called a polynomial function if it satisfies

$$f(x) = a_n x^n + a_{n-1} x^{n-1} + ... + a_2 x^2 + a_1 x + a_0$$

for all arguments x, where n is a non-negative integer and $a_0, a_1, a_2, ... a_n$, are constant coefficients. For example, the function $f : R \to R$, taking real numbers to real numbers, defined by

$$f(x) = x^3 - x$$

is a polynomial function of one variable. Polynomial functions of multiple variables can also be defined, using polynomials in multiple indeterminates, as in

$$f(x, y) = 2x^3 + 4x^2 y = xy^5 + y^2 - 7.$$

An example is also the function

$$f(x) = \cos(2 \arccos x)$$

which, although it doesn't look like a polynomial, is a polynomial function on $[-1, 1]$ since for every x from $[-1, 1]$ it is true that

$$f(x) = 2x^2 - 1$$

(see Chebyshev polynomials). Polynomial functions are a class of functions having many important properties. They are all continuous, smooth, entire, computable, etc

Orthogonality By using integral calculus. it is common to use the following to define the inner product of two functions f and g:

$$\langle f, g \rangle_w = \int_a^b f(x)g(x)w(x)dx$$

Here we introduce a nonnegative weight function $w(x)$ in the definition of this inner product. In simple cases, $w(x) = 1$, exactly.

We say that these functions are orthogonal if their inner product is zero:

$$\langle f, g \rangle_w = \int_a^b f(x)g(x)w(x)dx = 0$$

Hence the members of a set of functions $\{f_i, i = 1, 2, 3, ...\}$ are:

- orthogonal on the closed interval $[a, b]$ if

$$\langle f_i, f_j \rangle_w = \int_a^b f_i(x)f_j(x)w(x)dx = ||f_i||^2 \delta_{i,j} = ||f_j||^2 \delta_{i,j}$$

- orthonormal on the interval $[a, b]$ if

$$\langle f_i, f_j \rangle_w = \int_a^b f_i(x)f_j(x)w(x)dx = \delta_{i,j}$$

   where

$$\delta_{i,j} = \begin{cases} 1 & if\, i = j \\ 0 & if\, i \neq j \end{cases}$$

   is the "Kronecker delta" function.

1.3   Definitions of Orthogonal Polynomials

A system of polynomials $\{P_n\}$ which satisfy the condition of orthogonality

$$\int_a^b P_n(x)P_m(x)\omega(x)dx = 0, \qquad n \neq m$$

whereby the degree of every polynomial P_n is equal to its index n, and the weight function (weight) $\omega(x)$ on the interval (a, b) or (when a and b are finite) on $[a, b]$. Orthogonal polynomials are said to be orthonormalized, and are denoted by $\{\hat{P}_n\}$, if every polynomial has positive leading coefficient and if the normalizing condition

$$\int_a^b \hat{P}_n^2(x)\omega(x)dx = 1$$

is fulfilled.

1.4 Applications

Orthogonal polynomials have very useful properties in the solution of mathematical and physical problems. Orthogonal polynomials provide a natural way to solve, expand, and interpret solutions to many types of important differential equations.

Applications of orthogonal polynomials are found in varying degrees in

- Quadrature

- Numerical Interpolation

- Estimation of Error in Padé approximations

- Solutions to ordinary and partial differential equations

- Rational approximations and equilibrium distributions

- Factorization of second order difference equations

- Gauss quadrature for analytic functions

- Electrostatic interpretation of zeros of functins

-

Chapter 2

The Laguerre Polynomials

2.1 Introduction

The Laguerre polynomials, named after Edmond Laguerre $(1834 - 1886)$, are solutions of Laguerre's equation:

$xy + (1 - x)y + ny = 0$

2.2 Recursive relation, closed form, and generating function

One can also define the Laguerre polynomials recursively, defining the first two polynomials as

$L_0(x) = 1$

$L_1(x) = 1 - x$

and then using the following recurrence relation for any $k \geq 1$:

$$L_{k+1}(x) = \frac{1}{k+1}((2k + 1 - x)L_k(x) - kL_{k-1}(x))$$

The closed form is

The generating function for them likewise follows,

2.3 The first few terms of Laguerre polynomials

n	$L_n(x)$
0	1
1	$-x + 1$
2	$\frac{1}{2}(x^2 - 4x + 2)$
3	$\frac{1}{6}(-x^3 + 9x^2 - 18x + 6)$
4	$\frac{1}{24}(x^4 - 16x^3 + 72x^2 - 96x + 24)$
5	$\frac{1}{120}(-x^5 + 25x^4 - 200x^3 + 600x^2 - 600x + 120)$
6	$\frac{1}{720}(x^6 - 36x^5 + 450x^4 - 2400x^3 + 5400x^2 - 4320x + 720)$

Chapter 3

Zeros of Laguerre Polynomials

3.1 Introduction

3.2 Bounds for extreme zeros of some classical orthogonal polynomials

The following result is from
"Bounds for extreme zeros of some classical orthogonal polynomials" by K Driver, Department of Mathematics and Applied Mathematics, University of Cape Town, Private Bag X3, Ronde-bosch 7701, Cape Town, South Africa and K Jordaan, Department of Mathematics and Applied Mathematics, University of Pretoria, Pretoria, 0002, South Africa

Theorem 3.2.1. *Let $\{p_n\}_{n=0}^{\infty}$ be any sequence of polynomials orthogonal on the (finite or infinite) interval (c, d). Let g_{n-k} be any polynomial of degree $n - k$ that satisfies, for each $n \in N$, and $k \in \{1, \ldots, n-1\}$,*

$$f(x)g_{n-k}(x) = G_k(x)p_n(x) + H(x)p_{n+1}(x) \tag{3.2.1}$$

where $f(x) \neq 0$ for $x \in (c, d)$ and $H(x)$, $G_k(x)$ are polynomials with $\deg(G_k) = k$. Then, for each fixed $n \in N$,

1. *the n real, simple zeros of $G_k g_{n-k}$ interlace with the zeros of p_{n+1} if g_{n-k} and p_{n+1} are co-prime;*

2. *if g_{n-k} and p_{n+1} are not co-prime and have r common zeros, counting multiplicity, then*

   (a) *$r \leq \min\{k, n-k\}$;*

   (b) *these r common zeros are simple zeros of G_k;*

   (c) *no two successive zeros of p_{n+1} , nor its largest or smallest zero, can also be zeros of g_{n-k};*

   (d) *the $n - 2r$ zeros of $G_k g_{n-k}$, none of which is also a zero of p_{n+1}, together with the r common zeros of g_{n-k} and p_{n+1}, interlace with the $n+1-r$ remaining (non-common) zeros of p_{n+1}.*

Proof. Let $w_{n+1} < \ldots < w_1$ denote the zeros of p_{n+1}.

1 From (2), provided $p_{n+1}(x) \neq 0$, we have

$$\frac{f(x)g_{n-k}(x)}{p_{n+1}(x)} = H(x) + \frac{G_k(x)p_n(x)}{p_{n+1}(x)}. \tag{3.2.2}$$

Further,

$$\frac{p_n(x)}{p_{n+1}(x)} = \sum_{j=1}^{n+1} \frac{A_j}{x - w_j}$$

where $A_j > 0$ for each $j \in \{1, \ldots, n+1\}$ (cf. [22, Theorem 3.3.5]). Therefore (3) can be written as

$$\frac{f(x)g_{n-k}(x)}{p_{n+1}(x)} = H(x) + \sum_{j=1}^{n+1} \frac{G_k(x)A_j}{x - w_j}, x \neq w_j \tag{4} \tag{3.2.3}$$

Since p_{n+1} and p_n are always co-prime while p_{n+1} and g_{n-k} are co-prime by assumption, it follows from (2) that $G_k(w_j) \neq 0$ for any $j \in \{1, 2, \ldots, n+1\}$. Suppose that G_k does not change sign in the interval $I_j = (w_{j+1}, w_j)$ where $j \in \{1, 2, \ldots, n\}$. Since $A_j > 0$ and the polynomial H is bounded on I_j while the right hand side of (4) takes arbitrarily large positive and negative values on I_j, it follows that g_{n-k} must have an odd number of zeros in each interval in which G_k does not change sign. Since G_k is of degree k, there are at least $n - k$ intervals (w_{j+1}, w_j), $j \in \{1, \ldots, n\}$ in which G_k does not change sign and so each of these intervals must contain exactly one of the $n - k$ real, simple zeros of g_{n-k}. We deduce that the k zeros of G_k are real and simple and, together with the zeros of g_{n-k}, interlace with the $n + 1$ zeros of p_{n+1}.

2 If r is the total number of common zeros of p_{n+1} and g_{n-k} counting multiplicity then each of these r zeros is a simple zero of p_{n+1} and it follows from (2) that any common zeros of g_{n-k} and p_{n+1} must also be zeros of G_k since p_n and p_{n+1} are co-prime. Therefore, $r \leq min\{k, n-k\}$ and there must be at least $(n-2r)$ open intervals of the form $I_j = (w_{j+1}, w_j)$ with endpoints at successive zeros of p_{n+1} where neither w_{j+1} nor w_j is a zero of g_{n-k} or $G_k(x)$.

If G_k does not change sign in an interval $I_j = (w_{j+1}, w_j)$, it follows from (4), since $A_j > 0$ and H is bounded while the right hand side takes arbitrarily large positive and negative values for $x \in I_j$, that g_{n-k} must have an odd number of zeros in that interval. Therefore, in at least $(n-2r)$ intervals I_j either g_{n-k} or G_k, but not both, must have an odd number of zeros counting multiplicity. On the other hand, g_{n-k} and G_k have at most $(n - k - r)$ and $(k - r)$ real zeros respectively that are not zeros of p_{n+1}. We deduce that there must be at most $(n - 2r)$ intervals $I_j = (w_{j+1}, w_j)$ with endpoints at successive zeros w_{j+1} and w_j of p_{n+1} neither of which is a zero of $g_n - k$. It is straightforward to check that if the number of intervals $I_j = (w_{j+1}, w_j)$ with endpoints at successive zeros of p_{n+1} neither of which is a zero of g_{n-k} is equal to $n - 2r$, this is only possible if no two consecutive zeros of p_{n+1}, nor the largest or smallest zero of p_{n+1}, can be common zero(s) of p_{n+1} and g_{n-k}. This proves $a)$ to $c)$ and $d)$ follows from $c)$.

$\square$

3.3 Bounds for zeros of the Laguerre polynomials

The following is extracted from *"Ilia Krasikov, Bounds for zeros of the Laguerre polynomials, Journal of Approximation Theory 121 (2003) 287-291; http://www.elsevier.com/locate/jat*

Theorem 3.3.1. *Let x_1 and x_k be the least and the largest zeros of $L_k^{(\alpha)}(x)$, respectively. For $k \geq 7$, $\alpha \geq 8$ the following inequalities hold:*

$$x_1 > s - r + \frac{(s-r)^{2/3}}{2r^{1/3}}, \tag{3.3.1}$$

$$x_k < s + r + \frac{(s+r)^{2/3}}{2r^{1/3}}, \tag{3.3.2}$$

where

$$s = 2k + \alpha + 1, r = \sqrt{4k^2 + (2k-1)(2k+2)}$$

More precisely, all the zeros of $L_k^{(\alpha)}(x)$ are confined between the only two real roots of the following equation:

$$x^2 - 2sx + b^2 - 1) - 4sx^3 + 9s^2x^2 + (b^2 - 1)(b^2 - 1 - 6sx) = 0 \tag{3.3.3}$$

where $b = \alpha + 2$

It looks plausible that, up to the factor $\frac{1}{2}$, Theorem 1 gives the correct value of the second term of the corresponding asymptotics.

Proof. A real entire function $\phi(x)$ is in the Laguerre-Polya class $L - P$ if it has a representation of the form

$$\phi(x) = cx^m e^{-\alpha x^2 + \beta x} \prod_{k=1}^{\omega} \left(1 + \frac{x}{x_k}\right) e^{-x/x_k} \qquad (\omega \leq \infty,$$

where c, β, x_k are real, $\alpha \geq 0$, m is a nonnegative integer and $\sum x_k^{-2} < \infty$. Our main tool will be the following inequality valid for any $f \in L - P$ [7,9,10],

$$V_m(f(x)) = \sum_{j=-m}^{m} (-1)^{m+j} \frac{f^{(m-j)}(x)f^{(m+j)}(x)}{(m-j)!(m+j)!} \geq 0, \qquad m = 0, 1, \dots. \tag{3.3.4}$$

We will use $m = 2$ and set

$$V = 12V_2(y) = 3y''^2 - 4y'y''' + yy^{(4)}.$$

Notice that in our case, the positivity (and a plausible connection with the potential theory) can be seen directly by $V = \sum_{i \neq j}(x - x_i)^{-2}(x - x_j)^{-2}$, where $x_1, x_2, \dots$ are the zeros of y [3].

In the sequel, we deal with the function $t = t(x) = y'/y$, and set b, r, and s as in Theorem 1 to simplify some expressions. We also assume $x > 0$. Using differential equation (1) recursively to express the higher derivatives in V through y and y' we get

$$\frac{2x^3}{y^2}V = At^2 + 2Bt + C, \tag{3.3.5}$$

9

where

$$A = -2x(x^2 - 2sx + (b-1)(b+3)),$$

$$B = x^3 - (2s + b - 1)x^2 + (2bs - 3s + (b-1)(b+3))x + b - b^3,$$

$$C = (b - s - 1)(x^2 - 2sx - x + b^2 + b)$$

Observe that A is positive only for x in the interval (x_m^a, x_M^a),

$$x_{m,M}^a = 2k + \alpha + 1 \pm 2\sqrt{k^2 + \alpha k + k - \alpha - 1}.$$

Let also $x_m^c < x_M^c$ be the roots of C.

For the discriminant of the equation $At + 2Bt + C = 0$; in t we get

$$\Delta(x) = B^2 - AC = (x^2 - 2sx + b^2 - 1)^3 - 4sx^3 + 9s^2x^2 + (b^2 - 1)(b^2 - 1 - 6sx) \quad (3.3.6)$$

that is exactly expression (6).

We now state two Lemmas that are important for the theorem. theorem Lemma 1.
The equation $\Delta 9x) = 0$ has exactly two real roots $x_m^* < x_M^*$, provided $k \geq 2$ and $\alpha > -1$.
Moreover, $x_m^* < x_m^a < x_m^* < x_M^* < x_M^a < x_M^*$, if $k \geq 7$, $\alpha \geq 8$. end theorem (See the full
proof of this lemma in "Ilia Krasikov, Bounds for zeros of the Laguerre polynomials, Journal of
Approximation Theory 121 (2003) 287-291; http://www.elsevier.com/locate/jat) begin theorem
Lemma 2.
For $k \geq 7$, $\alpha \geq 8$, all the zeros of $L_k^{(\alpha)}$ are confined in (x_m^*, x_M^*) between the only two real roots
of the equation $\Delta(x) = 0$ end theorem

(See the complete proof of this lemma in "Ilia Krasikov, Bounds for zeros of the Laguerre
polynomials, Journal of Approximation Theory 121 (2003) 287-291; http://www.elsevier.com/locate/jat)
Proof of Theorem.
By the previous lemma it is enough to show that inequalities (4) and (5) hold for x_m and x_M
respectively. Since

$$s - r + \frac{(s-r)^{2/3}}{2r^{1/3}} < s < s + r + \frac{(s+r)^{2/3}}{2r^{1/3}},$$

and

$$\Delta(s) = -(s^2 - b^2 + 1)((s^2 - b^2)^2 - 3s^2 - b^2) < 0,$$

to prove (4) we just check

$$\Delta\left(s \pm r + \frac{(s \pm r)^{2/3}}{2r^{1/3}}\right) > 0.$$

Calculations yield

$$\frac{64r^6}{q^4}\Delta\left(s - r + \frac{(s-r)^{2/3}}{2r^{1/3}}\right) = q^8 - 32q^5r - 60q^6r^{4/3} - 32q^2r^2 + 96q^3r^{4/3} + 240q^4r^{8/3} + 144r^10/3 + 192qr^11/3,$$

where Since, as it is easy to check, we convince that the above expression is positive. Hence (4)
follows. The proof of (5) is similar using $\qquad \square$

10

3.4 Limit relation for the complex zeros of Laguerre polynomials

The following result is by *Mark V. DeFazio, Dharma P. Gupta, Martin E. Muldoon, on "Limit relations for the complex zeros of Laguerre and q-Laguerre polynomials", in J. Math. Anal. Appl. 334 (2007) 977-982; www.elsevier.com/locate/jmaa*

Theorem

Let $x_1(\alpha), \ldots, x_m(\alpha)$ be the m, $(2 \le m \le n)$ zeros of $L_n^{(\alpha)}(x)$ in a neighbourhood of $x = 0$ for $\alpha \sim -m$. Then

$$\lim \alpha \to -m \sum_{k=1}^{m} \frac{1}{x_k(\alpha)} = \frac{m(m-2n-1)}{m^2-1}. \tag{3.4.1}$$

Proof. Let α be close to $-m$. Let us number the zeros of $L^{(\alpha)}_n(x)$ so that $x_1, \ldots, x_m$ are near zero and $x_{m+1}, \ldots, x_n$ are near the positive real zeros of $L^{(m)}_{n-m}(x)$. The explicit formula (2) yields

$$\sum_{k=1}^{n} \frac{1}{x_k(\alpha)} = \frac{n}{\alpha+1}, \qquad \alpha \ne -1, \ldots, -n. \tag{3.4.2}$$

This is readily seen by considering the factorization

$$(n+\alpha n)^{-1} L_n^{(\alpha)}(x) = \left(1 - \frac{x}{x_1(\alpha)}\right) \cdots \left(1 - \frac{x}{x_n(\alpha)}\right) = 1 - x \sum_{k=1}^{n} \frac{1}{x_k(\alpha)} + \ldots, \tag{3.4.3}$$

and comparing with the expansion (2):

$$(n+\alpha n)^{-1} L_n^{(\alpha)}(x) = 1 - \frac{n}{\alpha+1} x + \ldots. \tag{3.4.4}$$

Letting $\alpha \to -m$ gives

$$\frac{n}{1-m} = \lim_{\alpha \to -m} \sum_{k=1}^{n} \frac{1}{x_k(\alpha)} \qquad = \lim_{\alpha \to \infty} \sum_{k=1}^{m} \frac{1}{x_k(\alpha)} + \lim_{\alpha \to \infty} \sum_{k=m+1}^{n} \frac{1}{x_k(\alpha)},$$

where the zeros in the first sum are the ones in the neighbourhood of 0. But the zeros in the second sum on the right approach those of $L^{(m)}_{n-m}(x)$ and hence, using (8), we get

$$\frac{n}{1-m} = \lim_{\alpha \to -m} \sum_{k=1}^{m} \frac{1}{x_k(\alpha)} + fracn - m1 + m, \tag{3.4.5}$$

which gives (7).

3.5 On the Asymptotic Distribution of the Zeros of Laguerre Polynomials

The following result is by WOLFGANG GAWKONSKI on "On the Asymptotic Distribution of the Zeros of Hermite, Laguerre, and Jonquière Polynomials" in JOURNAL OF APPROXIMATION THEORY 50. 2 14-23 i (1987)

11

Theorem

If $L_n^{(\alpha)}(z)$ denotes the Laguerre polynomial of degree n, $\alpha > -1$, and $N_n(\xi)$ is the number of zeros of $L_n^{(\alpha)}(z)$ not exceeding ξ, $\xi > 0$, then

$$\lim_{n\to\infty} \frac{1}{n} N_n(4n\xi) = \frac{2}{\pi} \int_0^\xi t^{-1/2}(1-t)^{1/2} dt, \qquad 0 < \xi \le 1. \tag{3.5.1}$$

Theorem 1 gives the asymptotic number of zeros in intervals of the form $(4n\xi_1, 4n\xi_2]$ that is in intervals the length of which tends to infinity with n. In contrast to this result the question arises: "how many" zeros are located in a "fixed" interval, $(\xi_1, \xi_2]$ say? The precise answer is given by:

THEOREM 2

Under the assumptions and with the notations of Theorem 1 we have

$$\lim_{n\to\infty} \frac{1}{\sqrt{n}} N_n(\xi) = \frac{2}{\pi}\sqrt{\xi}, \qquad \xi > o. \tag{3.5.2}$$

3.6 Monotonicity of zeros of Laguerre polynomials

Denote by x_{nk}, $k = 1, \ldots, n$, the zeros of the Laguerre polynomial $L_n^\alpha(x)$. We establish monotonicity with respect to the parameter α of certain functions involving $x_{nk}(\alpha)$. As a consequence we obtain sharp upper bounds for the largest zero of $L_n^\alpha(x)$

Theorem 3.6.1. *For every $n \ge 2$ and each k, $k = 1, \ldots, n$, the quantities*

$$\frac{x_{nk}(\alpha) - (2n + \alpha - 1)}{\sqrt{2(n + \alpha - 1)}}$$

are increasing functions of α, for $\alpha \ge -1/(n-1)$. Moreover, when $k = 1$ the above function increases in the entire range $\alpha \in (-1, \infty)$.

For the proof of above, see ([8])

3.7 Zeros of linear combinations of Laguerre polynomials from different sequences

We consider linear combinations of Laguerre polynomials L_n^α of the form $R_n^{\alpha,t} = L_n^\alpha + aL_n^{\alpha+t}$ and $S_n^{\alpha,t} = L_n^\alpha + bL_{n-1}^{\alpha+t}$ where $\alpha > -1$, $t > 0$ and a, $b \ne 0$. We recall that the Laguerre polynomials (cf. [34]) are orthogonal with respect to the weight function $e^{-x}x^\alpha$, $\alpha > -1$ on the interval $(0, \infty)/$.

For $0 < t \le 2$, we give proofs ([23]) for the interlacing of the zeros of $R_n^{\alpha,t}$ and $S_n^{\alpha,t}$ with the zeros of L_n^α, $L_n^{\alpha+t}$, L_{n-1}^α and $L_{n-1}^{\alpha+t}$.

We will make use of two well known identities (cf. [1], 22.7.30 and 22.7.29)

$$L_n^\alpha = L_n^{\alpha+1} - L_{n-1}^{\alpha+1} \tag{3.7.1}$$

$$xL_n^{\alpha+1}(x) = (x - n)L_n^\alpha + (\alpha + n)L_{n-1}^\alpha. \tag{3.7.2}$$

3.7.1 Linear combinations of Laguerre polynomials of the same degree

Theorem ([23])
Let

$$R_n^{\alpha,t} = L_n^\alpha + aL_n^{\alpha+t}, \qquad \alpha > -1.$$

For $0 < t \leq 2$, the zeros of $R_n^{\alpha,t}$ interlace with the zeros of (i) L_n^α, $L_n^{\alpha+t}$.
Proof ([23])
We have from [[13], Theorem 2.3] that the zeros of L_n^α interlace with the zeros of $L_n^{\alpha+t}$ for $0 < t \leq 2$ which implies that L_n^α has a different sign at successive zeros of $L_n^{\alpha+t}$ and vice versa. Evaluating (3) at successive zeros x_i and x_{i+1} of L_n^α we obtain

$$R_n^{\alpha,t}(x_i)R_n^{\alpha,t}(x_{i+1} = a^2 L_n^{\alpha+t}(x_i)L_n^{\alpha+t}(x_{i+1}), \qquad i = 1, 2, \ldots, n-1 \qquad < 0 \, for \, all \, a \neq 0.$$

Therefore $R_n^{\alpha,t}$ has a different sign at successive zeros of L_n^α and so the zeros interlace. The same argument shows that the zeros of $R_n^{\alpha,t}$ interlace with those of $L_n^{\alpha+t}$ by evaluating (3) at successive zeros of $L_n^{\alpha+t}$.

3.7.2 Linear combinations of Laguerre polynomials of different degree

Next we consider linear combinations of the type

$$S_n^{\alpha,t} = L_n^\alpha + bL_{n-1}^{\alpha+t}, \qquad b \neq 0, \alpha > -1 \tag{3.7.3}$$

We will need information on the interlacing properties of the two polynomials L_n^α and $L_{n-1}^{\alpha+t}$ in the linear combination. Theorem ([23])
Let $\alpha > 1$ and let

$$0 < x_1 < x_2 < \ldots < x_n$$

be the zeros of L_n^α

$$0 < y_1 < y_2 < \ldots < y_{n-1}$$

be the zeros of L_{n-1}^α

$$0 < t_1 < t_2 < \ldots < t_{n-1}$$

be the zeros of $L_{n-1}^{\alpha+t}$ and

$$0 < X_1 < X_2 < \ldots < X_{n-1}$$

be the zeros of $L_{n-1}^{\alpha+2}$
where $0 < t < 2$. Then $0 < x_1 < y_1 < t_1 < X_1 < x_2 < \ldots < x_{n-1} < y_{n-1} < t_{n-1} < X_{n-1} < x_n$
Proof ([23])
A simple computation using (1) and (2) leads to

$$(\alpha+1)L_n^{\alpha+1}(x) = (\alpha+n+1)L_n^\alpha(x) + xL_{n-1}^{\alpha+2}. \tag{3.7.4}$$

Evaluating (7) at successive zeros x_k and x_{k+1} of $L_n^\alpha(x)$, we obtain

$$x_k x_{k+1} L_{n-1}^{\alpha+2}(x_k) L_{n-1}^{\alpha+2}(x_{k+1}) = (\alpha + 1)^2 L_n^{\alpha+1}(x_k) L_n^{\alpha+1}(x_{k+1}).$$

The expression on the right is negative since the zeros of L_n^α and $L_n^{\alpha+1}$ interlace (cf. [12], Theorem 2.3]) and therefore
$x_k < X_k < x_{k+1}$ for each fixed k, $k = 1, \ldots, n-1$.
The zeros of L_{n-1}^α increase as α increases (cf. [34], p. 122), hence $y_k < t_k < X_k$ for each fixed k, $k = 1, \ldots, n-1$
Finally, since the zeros of L_n^α and L_{n-1}^α separate each other, we know that
$x_k < y_k < x_{k+1}$ for each fixed k, $k = 1, \ldots, n-1$
and this completes the proof
Theorem ([23])
For $0 < t \leq 2$, the zeros of $S_n^{\alpha,t}$ interlace with the zeros of (i) L_n^α, (ii) $L_{n-1}^{\alpha+t}$.
Proof ([23])
We know from Theorem 3.1 that the zeros of L_n^α interlace with the zeros of $L_{n-1}^{\alpha+t}$ for $0 < t \leq 2$ which implies that $L_{n-1}^{\alpha+t}$ has a different sign at successive zeros of L_n^α and vice versa. Evaluating (6) at successive zeros x_i and x_{i+1} of L_n^α we obtain

$$S_n^{\alpha,t}(x_i) S_n^{\alpha,t}(x_{i+1}) = b^2 L_{n-1}^{\alpha+t}(x_i) L_{n-1}^{\alpha+t}(x_{i+1}, \qquad i = 1, 2, \ldots, n-1 \qquad < 0, \forall b \neq 0.$$

Therefore $S_n^{\alpha,t}$ has a different sign at successive zeros of L_n^α and so the zeros interlace. The same argument shows that the zeros of $S_n^{\alpha,t}$ interlace with those of $L_{n-1}^{\alpha+t}$ by evaluating (6) at successive zeros of $L_{n-1}^{\alpha+t}$.

3.8 Convexity of the extreme zeros of Laguerre polynomials

The convexity theorem as noted by Sturm in ([32] and cited by [7]), can be summarised as follows.
Theorem
Let $y''(t) + F(t)y(t) = 0$ be a second-order differential equation in normal form, where F is continuous in (a, b). Let $y(t)$ be a nontrivial solution in (a, b), and let $x_1 < \ldots < x_k < x_{k+1} < \ldots$ denote the consecutive zeros of $y(t)$ in (a, b). Then

1. if $F(t)$ is strictly increasing in (a, b), $x_{k+2} - x_{k+1} < x_{k+1} - x_k$,

2. if $F(t)$ is strictly decreasing in (a, b), $x_{k+2} - x_{k+1} > x_{k+1} - x_k$.

3. if there exists $M > 0$ such that $F(t) < M$ in (a, b) then

$$\Delta x_k \equiv x_{k+1} - x_k > \frac{\pi}{\sqrt{M}},$$

4. if there exists $m > 0$ such that $F(t) > m$ in (a, b) then

$$\Delta x_k < \frac{\pi}{\sqrt{m}},$$

14

We say that the zeros of y are concave (convex) on (a, b) for the first (second) case. Hence we present the following theorem by

Theorem

The zeros of $L_n^\alpha(t)$ on $(0, 1)$ are

1. all convex if $n > 0$ and $-1 < \alpha \le 3$

2. all convex if $\alpha > 3$ and $0 < n < \frac{\alpha+1}{\alpha-3}$

3. concave for $t < t_0$ and convex for $t > t_0$ when $\alpha > 3$, $n > \frac{\alpha+1}{\alpha-3}$ and t_0 is defined by (3).

Moreover, for the distance between consecutive zeros we have the general estimate

$$\Delta x_k > \frac{\pi\sqrt{2}}{\sqrt{2\alpha n + \alpha + 2n^2 + 2n + 1}} \qquad k = 1, \ldots, n-1 \qquad (3.8.1)$$

and also if $x_k > t_0$ then

$$\Delta x_k > \frac{\pi}{\sqrt{F(x_k)}} \qquad k = 1, \ldots, n-1 \qquad (3.8.2)$$

and

$$\Delta x_k > \frac{\pi}{\sqrt{F(x_{k+1})}} \qquad k = 1, \ldots, n-1 \qquad (3.8.3)$$

Conclusion where F is defined by (2).

Proof For $|\alpha| < 1$, $t_0 < 0$, hence $F(t)$ will be decreasing on $(0, \infty)$. When $\alpha \ge 1$, $F(t)$ is increasing on $(0, t_0)$ and decreasing on $(t0, 1)$. Let x_1 denote the smallest zero of $L - n^\alpha$, then we know that $x_1 > \frac{\alpha+1}{n}$ (see [19]). This implies that when $t_0 < \frac{\alpha+1}{n}$, $F(t)$ will be decreasing on the interval (x_1, ∞). An easy calculation shows that this condition is equivalent to either $\alpha \le 3$ or $\alpha > 3$ and $n < \frac{\alpha+1}{\alpha-3}$. The estimates on the distance Δx_k follow from Theorem 2.1(3),(4). The maximum of F is at t_0 and $F(t_0) > 0$, therefore we can take $F(t_0)$ as M to obtain (4). For (5) and (6), we use the fact that when $x_k > t_0$, F is monotone decreasing on (x_k, x_{k+1}). In fact, F is monotone decreasing on $(0, 1)$ and tends to $-1/4$ as $t \to \infty$, so there is exactly one point t_1 on (t_0, ∞), where F crosses the x-axis. The form of the differential equation implies that if $F(t) < 0$ and $y(t) > 0$, the graph will be concave up and similarly, if $y(t) < 0$, the graph will be concave down.

Hence there can be at most one zero of the Laguerre polynomial to the right of t_1. This means that $F(x_{n-1})$ is positive, but $F(x_n)$ may be negative and therefore the index in (6) only runs up to $n-2$.

Conclusion

Bibliography

[1] Abramowitz, M. and Stegun, I. A. (Eds.) (1972). Handbook of Mathematical Functions with Formulas, Graphs, and Mathematical Tables, 9th printing. New York: Dover, p. 19.

[2] Andrews, George E.; Askey, Richard; Roy, Ranjan (1999), Special functions, Encyclopedia of Mathematics and its Applications 71, Cambridge University Press, ISBN 978-0-521-62321-6, MR 1688958

[3] Askey, R. A.; Roy, R. (2010), "Gamma function", in Olver, Frank W. J.; Lozier, Daniel M.; Boisvert, Ronald F.; Clark, Charles W., NIST Handbook of Mathematical Functions, Cambridge University Press, ISBN 978-0521192255, MR 2723248

[4] AF Beardon, K Driver, K Jordaan (2011). Zeros of polynomials embedded in an orthogonal sequence. *Numer Algor.* 57, 399-403.

[5] AF Beardon (2011). The theorems of Stieltjes and Favard. *Lect. Notes Math.*, 11(1), 247-262.

[6] C de Boor & EB Saff (1986). Finite sequences of orthogonal polynomials connected by a Jacobi matrix. Linear Algebra Appl. 75, 43-55.

[7] A. Deano, A. Gil, J. Segura (2004). New inequalities from classical Sturm theorems. *J. Approx. Theory* 131, 208-243.

[8] Dimitar K. Dimitrov, Fernando R. Rafaeli (2009). Monotonicity of zeros of Laguerre polynomials, *Journal of Computational and Applied Mathematics* 233 699-702

[9] D K Dimitrov & G P Nikolov (2010). Sharp bounds for the extreme zeros of classical orthogonal polynomials. *J. Approx. Th.*, 162, 1793-1804.

[10] E van Doorn (1987). Representations and bounds for zeros of orthogonal polynomials and eigenvalues of signsymmetric tri-diagonal matrices. *J. Approx. Theory.* 51 254-266.

[11] K Driver & K Jordaan (2011). Stieltjes interlacing of zeros of Laguerre polynomials from different sequence. *Indagat. Math. New Ser.*, 21 204-211

[12] K. Driver, K. Jordaan (2007). Interlacing of zeros of shifted sequences of one-parameter orthogonal polynomials, Numer. Math. 107 (4) 615-624.

[13] K. Driver, K. Jordaan (2007). Interlacing of zeros of shifted sequences of one-parameter orthogonal polynomials, *Numer. Math.* 107 (4) 615-624.

[14] K Driver (2011). Interlacing of zeros of Gegenbauer polynomials of non-adjacent degree from different sequences. *Numer. Math.*, 10.1007/s00211-011-0407-y.

[15] Dunster, T. M. (2010), "Legendre and Related Functions", in Olver, Frank W. J.; Lozier, Daniel M.; Boisvert, Ronald F.; Clark, Charles W., NIST Handbook of Mathematical Functions, Cambridge University Press, ISBN 978-0521192255, MR 2723248

[16] A Elbert (2001). Some recent results on the zeros of Bessel functions and orthogonal polynomials. *J. Comput. Appl.Math.* 133 6583.

[17] PC Gibson (2000). Common zeros of two polynomials in an orthogonal sequence. *J. Approx. Theory*, 105, 129-132.

[18] D E Gupta & M E Muldoon (2007). Inequalities for the smallest zeros of Laguerre polynomials and their q-analogues. *Journal of Inequalities in Pure and Applied Mathematics*, 8.

[19] W. Hahn (1933). Bericht Â¨uber die Nullstellen der Laguerrschen und der Hermiteschen Polynome. Jahresbericht der Deutschen Mathematiker- Vereinigung 44, 215-236.

[20] Havil, J. Gamma (2003). Exploring Euler's Constant. Princeton, NJ: Princeton University Press.

[21] M E H Ismail & M E Muldoon (1995). Bounds for the small real and purely imaginary zeros of Bessel and related functions. *Methods and Applic. of Analysis*, 2(1), 1-21.

[22] M E H Ismail & X Li (1992). Bounds on the extreme zeros of orthogonal polynomials. Proc. Amer. Math. Soc. 115 131140.

[23] Kathy Driver & Kerstin Jordaan (2009). Zeros of linear combinations of Laguerre polynomials from different sequences, *Journal of Computational and Applied Mathematics* 233 719-722

[24] Koornwinder, Tom H.; Wong, Roderick S. C.; Koekoek, Roelof; Swarttouw, RenÃ© F. (2010), "Orthogonal Polynomials", in Olver, Frank W. J.; Lozier, Daniel M.; Boisvert, Ronald F.; Clark, Charles W., NIST Handbook of Mathematical Functions, Cambridge University Press, ISBN 978-0521192255, MR 2723248

[25] I Krasikov (2003). Bounds for zeros of the Laguerre polynomials. *J. Approx. Theory*, 121, 287291.

[26] A Markov (1886). Sur les racines de certaines Â´equations (seconde note). *Math. Ann.* 27 177182.

[27] M E Muldoon (1993). Properties of zeros of orthogonal polynomials and related functions. *J. Comput. Appl. Math.* 48 167186.

[28] E R Neumann (1921). Zur Theorie der Laguerreschen Polynome. Jahresber. d. S.M.V. 30. 15.

[29] P. NEVAI (1979). "Orthogonal polynomials." *Mem. Amer. Math. Sot.* Vol. 18, No. 213. Amer. Math. Sot. Providence, R.I.

[30] N Obrechkoff (1963). Verteilung und Berechnung der Nullstellen reeler Polynome. VEB, Deutscher Verlag der Wissenschaften, Berlin.

[31] J Segura (2008). Interlacing of the zeros of contiguous hypergeometric functions. *Numer. Algor.*, 49, 387- 407.

[32] C. Sturm (1836). Memoire sur les 'equations diff'erentielles du second ordre. *J. Math. Pures Appl.*, 1 106-186.

[33] P.K. Suetin (2001), "Jacobi polynomials", in Hazewinkel, Michiel, *Encyclopedia of Mathematics, Springer*, ISBN 978-1-55608-010-4

[34] Szegő, Gábor (1939). Orthogonal Polynomials. Colloquium Publications. XXIII. *American Mathematical Society*. ISBN 978-0-8218-1023-1. MR 0372517

[35] L Vinet & A Zhedanov (2004). A characterization of classical and semiclassical orthogonal polynomials from their dual polynomials. *J. Comput. Appl. Math.* 172, 41-48.

[36] H S Wall & M Wetzel (1994). Quadratic forms and convergence regions for continued fractions. *Duke Math. J.*, 11, 983-1000